Usborne Activities

Sticker Dolly Dressing Fairies

Designed and illustrated by
Stella Baggott and Vici Leyhane
Written by
Leonie Pratt

Contents

The back cover folds out so you can "park" spare stickers there while you dress the dolls.

Meet the fairies

Hazel, Rosie and Willow are three magical fairy friends. They fly around together, looking after the trees in the woodlands and the flowers that grow in a meadow.

This is Hazel. She's always on the move, fluttering among the flowers and dancing through the treetops.

Rosie loves spending time with the flowers, too. When she thinks no one's listening, she sings magical songs to help them grow.
Willow likes chatting with the bugs and butterflies. She loves hearing about what's happening in the woods and meadows.

Treetop House

The fairies live together in Treetop House, at the top of a very tall oak tree. Every morning they are woken by the sunlight twinkling through the leaves.

By the Waterfall

It's a hot, sunny day and Hazel, Rosie and Willow have flown to a waterfall to cool down. They dance around the water lilies and splash about with their fishy friends.

Flower garden

In a corner of the meadow, there's a special flower garden that the fairies love to visit. Every day they sprinkle magical water over the flowers to help them grow.

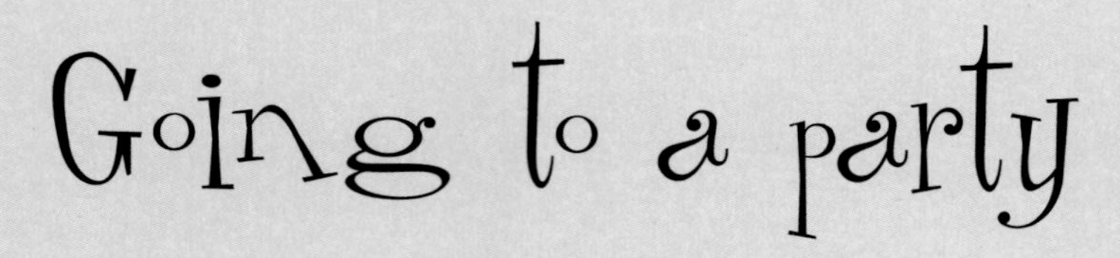

Going to a party

Carriages pulled by dragonflies have arrived to fly Hazel, Rosie and Willow to a birthday party. The fairies are wearing their best dresses and are bringing giftboxes filled with secret fairy surprises.

Rainy day

Rosie loves it when it rains. She puts on her raincoat and boots and skips through the puddles.

Morning dew

Hazel is flying through the meadow collecting dewdrops in a special bag. With a dash of fairy dust, the dewdrops turn into a magic potion for mending butterflies' wings.

Flying in the moonlight

A beautiful big moon is lighting the night sky and Hazel, Rosie and Willow are playing hide-and-seek with some fireflies. The fairies' dresses make it easy for them to hide among the bright flowers.

Fairy tea party

The fairies are having a tea party in the meadow. Hazel and Rosie have baked yummy cakes and laid them out on an old tree stump. Willow has been busy flying around the woods, inviting all their friends to the party.

The fairies are flitting through the cornfields, collecting grains of corn and juicy berries for their store cupboard. Willow likes stopping to talk to the friendly field mice who make their nests at the edge of the field.

Magical palace

Hazel, Rosie and Willow are going to visit the Fairy Queen, who lives in a magical palace deep in the meadow. The fairies want to look their best, so they are wearing their prettiest dresses and sparkly tiaras.

Chasing Snowflakes

It's a chilly winter morning and snowflakes are falling gently from the clouds. The fairies have wrapped up in coats, boots and hats, and have flown high into the sky to see how many snowflakes they can catch.

Fairy Wishes

The fairies are hiding in the strawberry patch. Anyone who finds them will be granted a special wish.

Series editor: Fiona Watt • Images of flowers and leaves on sticker pages © Alamy, © Digital Vision, © Digital Stock
This edition first published in 2015 by Usborne Publishing Ltd., 83-85 Saffron Hill, London, EC1N 8RT, England www.usborne.com This edition first published in America in 2015. U.E. Printed in China.